MASTER

XR

IPHONE XR USER GUIDE FOR BEGINNERS, NEW IPHONE XR USERS AND SENIORS (2019 VERSION)

Tech Reviewer

TABLE OF CONTENT

Introduction 13

iPhone XR Processor 14

iPhone XR Battery 15

iPhone XR Display 16

Absence of Home Button 17

Single Lens Portrait Mode 17

Symmetrical Speaker Ports 18

Exclusive Wallpapers 18

Lightweight Aluminum Frame Unlike the iPhone X, XS and XS Max, the XR has an aircraft-grade custom aluminum alloy with matte finishing. This makes the phone last longer and have light weight while showing off the beautiful colors. .. 19

Haptic Touch 19

Dust and Water Resistance 20

Getting Started: How to set up your iPhone XR .. 21

How to Use the Buttons and Sockets on the iPhone XR 26

List of Screen Icons 27

How to Insert SIM in iPhone XR 28

How to Charge the Battery for the iPhone XR 29

How to Extend the Device Battery Life................29

How to turn on iPhone XR................30

How to turn off iPhone XR................30

Going Home on your iPhone XR................31

How to Choose Ringtone on the iPhone XR........31

How to Choose Message Tone on the iPhone XR 32

How to Set/ Change Language on iPhone XR.....33

How to Use the Control Centre................34

How to Choose Settings for the Control Centre.34

How to set up Apple ID on iPhone XR................35

How to Set Up Apple Pay................37

How to check out with Apple Pay................38

How to use Siri on iPhone XR................39

CHAPTER 2: BASIC FUNCTIONS................43

How to Wake and Sleep Your iPhone XR...........43

How to Set up Face ID on iPhone XR................44

How to Unlock your iPhone XR using Face ID....45

How to make Purchases with Face ID on iPhone XR................47

How to Transfer Content to your iPhone XR from an Android Phone................47

How to Setup Vibration 48

How to Set Screen Brightness 49

How to Control Notification for Specific Apps ... 50

How to Control Group Notification 51

How to Set Do Not Disturb 51

How to Turn PIN on or Off 53

How to Change Device PIN 53

How to Unblock Your PIN 54

How to Confirm Software Version 55

How to Update Software 55

How to Control Flight Mode 55

How to Choose Night Shift Settings 56

How to Control Automatic Screen Activation 56

How to enable Location Services/ GPS on your iPhone XR 56

How to Turn off location services on iPhone selectively 57

How to Turn off location services on iPhone completely 58

How to Use Music Player 59

How to Navigate from the Notch 60

How to keep Track of documents 61

How to Move Between Apps 63

How to Force Close Apps in the iPhone XR 64

How to Arrange Home Screen Icons 65

Complete iPhone XR Reset Guide: How to perform a soft, hard, factory reset or master reset on the iPhone XR 65

How to Restart your/ Soft Reset iPhone 66

How to Hard Reset/ Force Restart an iPhone XR 67

How to Factory Reset your iPhone XR (Master Reset) 68

How to Use iTunes to Restore the iPhone XR to factory defaults 69

How to Choose Network Mode 70

How to set a reminder on iPhone XR 71

How to set a Recurring Reminder on your iPhone XR 73

How to get Battery Percentage on iPhone XR ... 74

How to take a Screenshot 74

CHAPTER 3: Calls and Contacts 76

How to Make Calls and Perform Other Features on Your iPhone XR 76

How to Call a Number 76

How to Answer Call 76

How to Control Call Waiting 77

How to Call Voicemail 77

How to Control Call Announcement 78

How to Add, Edit, and Delete Contacts on iPhone XR 79

How to Add Contacts 79

How to Save Your Voicemail Number 79

How to Merge Similar Contacts 80

How to Copy Contact from Social Media and Email Accounts 80

How to Create New Contacts from Messages On iPhone XR? 81

How to Add a Caller to your Contact 81

How to Add a contact after dialing the number with the keypad 82

How to Import Contacts 83

How to Delete contacts 84

How to Manage calls on your iPhone XR 84

How to Block Calls on the iPhone XR 85

How to Block Specific Numbers/Contacts on Your iPhone XR 86

How to Unblock Calls or Contacts on your iPhone XR 86

How to Use and Manage Call Forwarding on your iPhone XR 87

How to Cancel Call Forwarding on your iPhone XR 88

How to Manage Caller ID Settings and Call Logs on your iPhone XR 89

How to View and Reset Call Logs on your iPhone XR 89

How to Reset Call Logs 90

CHAPTER 4: Messages and Emails 91

How to Set up your Device for iMessaging 91

How to Compose and Send iMessage 91

How to Set up your Device for SMS 92

How to Compose and Send SMS 92

How to Set up Your Device for MMS 92

How to Compose and Send SMS 93

How to Hide Alerts in Message app on your iPhone XR 94

How to Set up Your Device for POP3 Email 94

How to Set up Your Device for IMAP Email 95

How to Set up Your Device for Exchange Email . 97

How to Create Default Email Account 98

How to Delete Email Account 98

How to Compose and Send Email 98

CHAPTER 5: Manage Applications and Data 100

How to Install Apps from App Store 100

How to Uninstall an App 100

How to Delete Apps Without Losing the App Data 101

How to Control Offload Unused Apps 101

How to Control Bluetooth 102

How to Control Automatic App Update 102

How to Choose Settings for Background Refresh of Apps 103

How to Synchronize using iCloud 105

How to manually add or remove music and videos to your iPhone XR 105

How to Choose Settings for Find my iPhone 106

How to Use Find My iPhone 106

How to Downgrade iOS System on Your iPhone 108

CHAPTER 6: Internet and Data 112

How to Set up your Device for Internet 112

How to Use Internet Browser 112

How to Clear Browser Data 113

How to Check Data Usage 113

How to Control Mobile Data 113

How to Control Data Roaming 114

How to Control Wi-fi Setup 114

How to Join a Wi-fi Network 114

How to use your iPhone as a Hotspot 115

How to Control Automatic Use of Mobile Data . 115

CHAPTER 7: What is iCloud Backup and How to Use it 116

What is iCloud Backup? 116

How to sign into iCloud on your iPhone XR 117

How to Sign Out of iCloud on Your iPhone XR .. 118

How to Use iCloud Backup 119

How to Troubleshoot if iCloud isn't Working 119

How to share a calendar on iPhone XR via iCloud 121

CHAPTER 8: Troubleshooting the iPhone XR Device 123

Troubleshooting Basic Functions 123

Calls and Voicemail Troubleshooting 130

Messages and Email Troubleshooting 132

Entertainment and Multimedia Troubleshooting 133
Connectivity Troubleshooting............................143
CHAPTER 9: Conclusion 150

How to Use this Book

Welcome! Thank you for purchasing this book and trusting us to lead you right in operating your new device. This book has covered every details and tips you need to know about the iPhone XR for to get the best from the device.

To better understand how the book is structured, I would advise you read from page to page after which you can then navigate to particular sections as well as make reference to a topic individually. This book has been written in the simplest form to ensure that every user understands and gets the best out of this book. The table of content is also well outlined to make it easy for you to reference topics as needed at the speed of light.

Thank you.

Other Books by Same Author

- Fire TV Stick; 2019 Complete User Guide to Master the Fire Stick, Install Kodi and Over 100 Tips and Tricks https://amzn.to/2FnmcQ9

- Mastering Your iPhone X: iPhone X User Guide for Beginners and Seniors (2019 Version) https://amzn.to/2J1ywGW

- Amazon Echo Dot 3rd Generation: Advanced User Guide to Master Your Device with Instructions, Tips and Tricks https://amzn.to/31PaBTF

Introduction

In 2018, Apple followed the footsteps of iPhone XS and XS Max to launch the iPhone XR which was tagged the "Cheapest iPhone device of the year." This alone was enough to make users excited about owning a new iPhone XR. However, it is important to note that the term "cheap" does not mean the device has a low standard. Even with its low pricing, the iPhone XR still comes packed with lots of features and abilities.

While the other devices were sold at retail prices of $1,099 and $999, the iPhone XR was launched at $749. This price may be high for android users, but old users of the iPhone saw this as a good deal.

Below are the prices of the iPhone XR with different storage capacities:

- $749 for 64 GB
- $799 for 128 GB
- $899 for 256 GB

The iPhone XR currently operates on iOS 12 which is familiar to many users as the operating system was launched even before the iPhone XR.

While you may think that the iPhone XR is different from the other devices released same year, be assured that the quality is the same as Apple did not lower their standard one bit in this device.

The iPhone XR is available in 6 colors: Yellow, Blue, Black, Coral, White and Red

Although the iPhone XR has more beauty than XS and XS Max, its camera and display feature is lower than that of the other devices. If you are currently using the iPhone 8 and wondering which device to switch to, the iPhone XR is still your best bet.

iPhone XR Processor

This device has same A12 bionic chip as the XS and XS Max. You may be wondering why the noise on this chip. When compared to all the processors

Apple has used from inception, the A12 bionic processor is still their best decision so far. This processor has several processing cores at extra high energy levels which allows you to carry out intensive tasks on your device. Apple says that the A12 bionic chip has been designed to power a minimum of 5 trillion operations per second.

iPhone XR Battery

The iPhone XR has a battery size of 2,942 mAh while the iPhone XS runs on 2,658 mAh. So, for less amount paid for the iPhone XR, you get to enjoy several features and functions.

According to reports by Apple, below are the iPhone XR battery specifications:

- Up to 65 hours audio playback on wireless
- Up to 25 hours talk time on wireless
- Up to 16 hours video playback on wireless
- Up to 15 hours internet use

Apple also gave the bonus of wireless charging on the iPhone XR. This device gives you an extra one and half hour battery life against the iPhone 8.

iPhone XR Display

Several users had anticipated that the XR would have the popular OLED display, however, Apple chose to go with the LCD screen on this one.

Although the LCD screen may be thought as something of the past, Apple prefers to see it as something from the future. With the LCD, the iPhone XR is the first device to have the entire front face of the camera covered. This is why the iPhone XR is tagged the "Liquid Retina Display." This is simply to say that Apple have removed the chin and forehead design that most users know for a long time.

The Liquid Retina Display is also Apple's first LCD device that has the Tap to Wake ability. It also supports the True Tone for adjustments based on lighting in your environment.

You may not consider the screen fancy as other devices are built with the amazing OLED display against this LCD display, however, if you are one for size, then the 6.1-inch screen size should impress you compared to 5.8 inches that we have on the iPhone X and XS. This is second to the iPhone XS Max that is 6.5 inches.

Absence of Home Button

Similar to the iPhone X, Apple company stopped the Home button, this means no more Touch ID.

The display on the iPhone XR stretches from the top to the bottom and from one edge to the other without any button on the front of the phone.

On the screen of your iPhone XR, you have the notch that contains the speaker, TrueDepth camera system and the sensors.

Single Lens Portrait Mode

One of the most used features of the iPhone X is the portrait mode photos and the dual camera system which are not present in the iPhone XR.

Apple only included the single lens assembly in the iPhone XR, however, you can replicate the Portrait mode effect by using software and upgraded Neural engine.

The iPhone XR uses the single camera to first detect a user's face before separating it from the background then it applies a soft bokeh effect to everything in the picture minus the main subject. With the Apple's depth control feature, you can further adjust it to fine tune the level of blur and bokeh.

Symmetrical Speaker Ports

The iPhone XR comes with symmetrical ports at the bottom of your iPhone. On each side of the lightning port, you have six circles that can serve as a microphone or speaker.

Exclusive Wallpapers

On each iPhone XR, you have the pre-loaded custom wallpapers that are designed to match with the exterior of the device. These wallpapers are beautiful and are excusive to the iPhone XR.

Lightweight Aluminum Frame

Unlike the iPhone X, XS and XS Max, the XR has an aircraft-grade custom aluminum alloy with matte finishing. This makes the phone last longer and have light weight while showing off the beautiful colors.

Haptic Touch

Apple has removed the 3D Touch on this device and replaced it with "Haptic Touch" which gives similar experience as the 3D touch. To use this feature, just press any applicable element for some seconds until you feel some vibration. With this, you have performed a similar 3D touch action.

There are some limitations on this like not being able to use Haptic Touch on Home Screen icons when you wish to view actions and widgets, but you can access widgets from the control center.

Apple has however promised to improve on this feature.

Dust and Water Resistance

The iPhone XR does not get affected by water and dust as the phone is built to withstand being in water up to one meter deep for as long as 30 minutes. However, Apple has warned that this is not permanent as normal wear and tear ca reduce the phone's ability to resist water and dust.

Other features of the iPhone XR include

- 12-MP camera, the True Depth camera system with Face ID
- Introduction of Memoji

- Wireless charging
- P3 wide color gamut
- True Tone
- Supports Dual SIM - one nano-SIM and one eSIM
- Bluetooth 5 as well as most recent updates in LTE.

CHAPTER 1

Getting Started: How to set up your iPhone XR

Setting up your device is the first and most important step to getting started with your iPhone XR. Follow these steps for a seamless experience.

1. Firstly, you need to power on your device. To do this, press and hold the side button. Now, you will see **"Hello"** in various languages. The screen would present options to set up your device. Follow the

options presented on the screen of your device.

Note: From the Hello screen, you can activate the **Voice Over or Zoom Option** which is helpful for the blind or those with low vision.

2. A prompt would come up next to select your language and country/ region. It is important you select the right information as this would affect how information like date and time etc. is presented on the device.

3. Next is to manually set up your iPhone XR by tapping "**Set up Manually**". You can choose the "**Quick Start**" option if you own another iOS 11 or later device by following the onscreen instruction. If you don't have this, then set up your iPhone manually.

4. Now, you have to connect your device to a cellular or Wi-Fi network or iTunes to

activate your phone and continue with the setup. You should have inserted the SIM card before turning on the phone if going with the cellular network option. To connect to a Wi-Fi network, just tap the name of your Wi-fi and it connects automatically if there is no password on the Wi-fi. If there is a security lock on the Wi-fi, the screen would prompt you for the password before it connects.

5. At this stage, you can turn on the Location services option to give access to apps like **Maps** and **Find my Friends**. This option can be turned off whenever you want. You would see how to turn on the location services and how to turn it off completely on your iPhone in a later part of this book.

6. Next is to set up your Face ID. The face ID feature gives you access to authorize purchases and unlock your devices. To setup the Face ID now, tap Continue and

follow the instructions on the screen. You can push this to a later time by selecting "**Set Up Later in Settings.**"

7. Whether you setup Face ID now or later, you would be required to create a four-digit passcode to safeguard your data. This passcode is needed to access Face ID and Apple Pay. Tap **"Passcode Option"** if you would rather set up a four-digit passcode, custom passcode or even no passcode.

8. If you have an existing iTunes or iCloud backup, or even an Android device, you can restore the backed-up data to your new phone or move data from the old phone to the new iPhone. To restore using iCloud, choose **"Restore from iCloud Backup"** or **"Restore from iTunes Backup"** to restore from iTunes to your new iPhone XR. In the absence of any backup or if this is your first device then select **"Set Up as New iPhone"**.

9. To continue, you would need to enter your Apple ID. If you have an existing Apple account, just enter the ID and password to sign in. In case you don't have an existing Apple ID or may have forgotten the login details, then select **Don't have an Apple ID or forget it.** If you belong to the class that have multiple Apple ID, then select **Use different Apple IDs for iCloud & iTunes** on the screen of the phone.
10. To proceed, you need to accept the iOS terms and conditions.
11. Next is to set up Siri and other services needed on your device. Siri needs to learn your voice so you would need to speak few words to Siri at this point. You can also set up the iCloud keychain and Apple Pay at this point.
12. Set up screen time. This would let you know the amount of time you spend on

your device. You can also set time limits for your daily app usage.

13. Now turn on automatic update and other important features.
14. Click on **"Get Started"** to complete the process. And now, you can explore and enjoy your device.

How to Use the Buttons and Sockets on the iPhone XR

Side button: This is the first button by the top right side of the phone formerly called the **"Sleep/ Wake"** button. You use this button to power on the device and turn on screen lock.

Silent Mode Key: This is the sliding key at the top left side of the device. You move this key either up or down to switch on or off the silent mode. When silent mode is on, you would not get any sound notification on your device.

Volume keys: These keys are used to increase or reduce volumes both on calls, when listening to music or to adjust the ring volume. You can also use it to mute an incoming call alert.

Camera lens: to take pictures or videos.

Lightning port: this is the socket at the bottom of the device used to charge the device or used to plug in headset for handsfree call or to listen to music.

List of Screen Icons

Wi-Fi icon: whenever you are connected to Wi-fi, you would see this icon as a notification.

Data Connection Icon: this notifies you of an active data connection.

Network Mode Icon: Here gives you information on the network mode of your phone at the time.

Flight Mode icon: is a notification that flight mode is activated.

Signal Strength icon: shows the strength of your connection per time.

Battery Charging icon: this shows that your battery is charging.

Battery Icon: shows the level of your battery, however it doesn't show the percentage level. We would talk on how to see the percentage level later in the book.

How to Insert SIM in iPhone XR

Before you can use your device, you must have inserted the SIM. To do this, please follow the steps below:

- The iPhone pack comes with an opener used to open the SIM and memory card holder.
- Put the opener into the tiny hole in the SIM holder.
- Pull out the SIM holder once the opener is able to clip it out.

- Set your SIM to ensure that the angled corner of the SIM is placed in the angled corner of the SIM holder. Note that iPhone XR only support Nano SIMs.
- Push in the SIM holder once the SIM has been correctly placed.
- Now your SIM is ready to be used.

How to Charge the Battery for the iPhone XR

It is important that you charge your phone often to ensure its ready for use at all times.

- Connect the phone charger to a power socket and then connect the USB side to the lightning port at the bottom of the phone.
- To know that your battery is charging, you would see the battery charging icon displayed at the top of the screen.
- At the top right side of the screen, you would see your battery level. The more

the colored section, the more power the device have and vice versa.

How to Extend the Device Battery Life

Some apps and services on the iPhone XR draw lots of power which would drain the battery life faster. You can turn on low power mode to reduce the power consumption.

- From **Settings**, go to **Battery**.
- Turn the switch beside **Low Power Mode** to the right to enable it.
- Go to Home screen.

How to turn on iPhone XR

The following steps would show how to turn on the iPhone XR:

- Press the **Side** button until the iPhone comes on.

- Once you see the Apple logo, release the button and allow your iPhone to reboot for about 30 seconds.
- Once the iPhone is up, you would be required to input your password if you have one.

How to turn off iPhone XR

Follow the steps below to turn off your iPhone XR:

- Hold both the volume down and the side button at same time.
- Release the buttons once you see the power off slider.
- Move the slider to the right for the phone to go off.
- You may also use the side and volume up button, only thing is, you may take a screenshot in error rather than shutting down the phone.

Going Home on your iPhone XR
- Regardless of where you are on your iPhone XR, to return to the home screen, simply swipe the screen from the bottom up.

How to Choose Ringtone on the iPhone XR
- From the Home screen, go to **Settings.**
- Click on **Sounds & Haptics.**
- Then click on **Ringtone.**
- You may click on each of the ringtones to play so you can choose the one you prefer.
- Select the one you like then click the "**< Back"** key at the top left of the screen.
- Slide the page from bottom up to return back to Home screen.

How to Choose Message Tone on the iPhone XR
- From the Home screen, go to **Settings.**

- Click on **Sounds & Haptics.**
- Then click on **"Text Tone".**
- You may click on each of the message tones to play so you can choose the one you prefer.
- Select the one you like then click the "**< Back**" key at the top left of the screen.
- Slide the page from bottom up to return back to Home screen.

How to Set/ Change Language on iPhone XR
- From the Home screen, click on the **Settings** option.
- Select **General** on the next screen.
- Then click on **Language and Region.**
- Click on **iPhone Language** to give you options of available languages.
- Choose your language from the drop-down list and tap **Done.**
- You would see a pop-up on the device screen to confirm your choice. Click on

Change to (Selected Language) and you are done!

How to Set/ Change Date/ Time on iPhone XR

- From the Home screen, click on the **Settings** option.
- Select **General** on the next screen.
- Then click on **Date & Time**.
- On the next screen, beside the **"Set Automatically"** option, move the switch right to turn it on.
- Slide the page from bottom up to return back to Home screen.

How to Use the Control Centre

- From the right side of the notch, swipe down to view the control center.
- Click on the needed function to either access it or turn it on or off.
- Move your finger up on the needed function to choose the required settings.
- Once done, return to home screen.

How to Choose Settings for the Control Centre

- Go to Settings> Control Centre.
- On the next screen, beside the **Access within Apps** option, move the switch to turn it on or off.
- Click on **Customize Controls.**
- For each function you want to remove, click on the minus (-) sign.
- To add icon under **More Controls,** click on the plus (+) sign at the left of each of the icons you want to add.
- Click on the move icon beside each function and drag the function to the desired position in the control centre.
- And you are done.

How to set up Apple ID on iPhone XR

- Go to the **Settings** option.
- At the top of your screen, click on **Sign in to your iPhone.**

- Choose **Don't have an Apple ID or forgot it?**
- A pop-up would appear on the screen and you click on **Create Apple ID.**
- Input your date of birth and click on Next.
- Then input your first name and last name then click on **Next**.
- The next screen would present you with the email address option. Click on **"Use your current email address"** if you want to use an existing email or click on **"Get a free iCloud email address"** if you want to create a new email.
- If using an existing email address, click on it and input your email address and password.
- If creating a new one, click on the option and put your preferred email and password.
- Verify the new password.

- Next option is to select 3 Security Questions from the list and provide answers.
- You have to agree to the device's Terms and conditions to proceed.
- Select either **Merge** or **Don't Merge** to sync the data saved on iCloud from reminders, Safari, calendars and contacts.
- Click on **OK** to confirm the **Find My iPhone is turned on.**

How to Set Up Apple Pay

- First is to add your card, either debit, credit or prepaid cards to your iPhone.
- To use the Apple Pay, your device should be updated to the latest iOS version.
- You should be signed into the iCloud using your Apple ID.
- To use the Apple Pay account on multiple devices, add your card to each of the devices.

To add your card to Apple Pay, do the following:

- Go to Wallet and click on
- Follow the instructions on the screen to add a new card. On iPhone XR, you can add as much as 12 cards. You may be asked to add cards that is linked to your iTunes, cards you have active on other devices or cards that you removed recently. Chose the cards that fall into the requested categories and then input the security code for each card. You may also need to download an app from your card issuer or bank to add your cards to the wallet.
- When you select **Next,** the information you inputted would go through your bank or card issuer to verify and confirm if the card can be used on Apple Pay. Your bank would contact you if they need further information to verify the card.
- After the card is verified, click Next to begin using Apple Pay.

How to check out with Apple Pay

Here are useful steps to check out on Apple Pay for your daily transactions:

- To make a payment at a checkout terminal, double-press the side button to open the Apple Pay screen.
- Look at the iPhone screen to verify with Face ID (or enter your passcode).
- And then place the iPhone XR near the payment terminal.
- If you're using Apple Pay Cash, double-press the side button to approve the payment.

How to use Siri on iPhone XR

Apples' virtual assistant is a delight to work with, everyone loves Siri, and most of the time you spend with her involves getting an answer, but she can do more than answer questions.

1. **How to Set up Siri on iPhone XR**

To use Siri on your iPhone XR, you have to set it up like you set up the Face ID. Find below the steps to do this on your iPhone XR.

- Click on **Siri & Search** from the **Settings** app.
- Beside the option "**Press Side Button for Siri**", move the switch to the right to enable the function.
- A pop-up notification would appear on the screen, select "**Enable Siri**".
- Switch on the "**Listen to Hey Siri**" option and follow the instructions you see on the screen of your iPhone. (To use Siri when your phone is locked, activate the **Allow Siri When Locked** option).
- Click on **language** and select the desired language.
- Click on the < **Siri & Search** button at the top left of the screen to go back.
- Scroll and select **Siri Voice**.

- On the next screen, select accent and gender.
- Click on the **< Siri & Search** button at the top left of the screen to go back.
- Select **"Voice Feedback"**.
- Choose your preferred setting.
- Click on **< Siri & Search** at the top left of the screen.
- Select **My Information.**
- Click on the contact of choice. If you set yourself as the owner of the phone, the device would use your data for various voice control functions like navigating home. You can create yourself on the contact by following the steps given in creating contact.
- Select the desired application.
- Next to **"Search and Siri Suggestions"** slight left or right to turn on or off.

Now Siri is set up and ready to be used.

2. **How to Activate Siri on the iPhone XR**

There are 2 ways to activate Siri on your iPhone XR.

- Voice option. If you enabled "Hey Siri", then you can begin by saying "Hey Siri" and then ask Siri any question.
- Using the side button. To wake Siri, press the side button and ask your questions. Once you release the side button, Siri stops listening.

3. **How to Exit Siri**

To exit Siri, follow the simple step below.

- Press the side button or swipe u from the bottom of the display to exit Siri.

CHAPTER 2: BASIC FUNCTIONS

How to Wake and Sleep Your iPhone XR

Waking and sleeping your iPhone XR will preserve your battery Life and make it long lasting; here are the steps for wake and put your iPhone XR to sleep.

- There are two ways to wake your iPhone XR and to see the lock screen; either tap the screen or just picking up the device and glancing at it can wake the iPhone XR.
- From the lock screen, you can use either the camera or the flashlight by a simple gesture. Tap and hold the individual icon until you hear a click sound.
- Simply press the side button to make the iPhone XR go to sleep. With the Apple Leather Folio case designed for the iPhone XR, simply open the case to wake and close the case to sleep your device.

How to Set up Face ID on iPhone XR

There are several things you can do just by glancing at your device. With Face ID, you can unlock your device, sign into apps, authorize purchases and lots more. Before setting up the Face ID, ensure nothing is covering your face or the TrueDepth camera. Apple designed the Face ID to work well with contacts and glasses. To get the best result, let your iPhone be about an arm's length from your face (10 – 20 inches or closer). Now, see steps to set up Face ID:

- Visit Settings > Face ID & Passcode. Input your passcode if asked.
- Select **Set Up Face ID**.
- Put your device in portrait orientation and place your face in front of your iPhone and tap "Get Started"
- The screen comes up with a frame, set your face to fit into the frame and move your head slowly until the circle is

complete. Tap **Accessibility Options** if you are not able to move your head.
- Once done with the first Face scan, click on continue.
- Move your head slowly again until the circle is completed for the second time.
- Tap **Done** to complete.
- If you do not already have a passcode set, you would be prompted to create one as an alternative option for identity verification.
- Go to **Settings** then **Face ID & Passcode** to activate features to go with the Face ID. This includes iTunes & App Store; iPhone Unlock and Safari AutoFill.

How to Unlock your iPhone XR using Face ID

The wonder of the iPhone XR is the ability to unlock the device with Face ID. Follow the steps to unlock your iPhone.
- Go to **Settings** then **Face ID & Passcode.**

- Go to the option **"Use Face ID For"** and switch on **iPhone Unlock.**
- To unlock your iPhone, wake the device first then you glance into the screen.
- The iPhone XR would automatically scan your face and authorize the login attempt.
- Once successful, the **lock icon** on the phone screen will open.
- To unlock the device, simply swipe from the bottom of the iPhone up to show the home screen.

With the iPhone XR, only a single face is supported on Face ID unlike the Touch ID on other iPhones. You would be unable to create faces of friends and family members on your iPhone. The only way a third party can access your phone is to manually input the password. This option would also work for you if the Face ID isn't working.

How to make Purchases with Face ID on iPhone XR

If the iTunes and App Store are activated for Face ID under **Face ID & Passcode,** you can use the Face ID to carry out purchases on the App store, iTunes Store and iBooks store.

Follow the steps below.
1. Open any of the stores on your phone.
2. Search for the items you want to purchase and click on it.
3. To make payment, click the **Side** button twice and look at your iPhone XR.
4. Once its completed, a message would pop up showing **Done** with a **Checkmark** on your device screen.

How to Transfer Content to your iPhone XR from an Android Phone

You can move contents to your device from an Android mobile phone when you first activate the device or after you did a factory reset. To do this, you would see the **Apps and Data** option on your screen.

- Under **Apps and Data,** click on "**Move Data from Android**".
- You have to install the app "**Move to iOS**" on the android phone before you can move data.
- Click on **Continue** when you have downloaded the app.
- Follow the instructions you see on the screen to move data from the Android to the iPhone XR.

How to Setup Vibration

- From the Home screen, go to **Settings**.
- Click on **Sounds & Haptics.**
- Toggle the switch next to "**Vibrate to Ring**" to enable or disable vibration when the silent mode is disabled.

- Toggle the switch next to **"Vibrate on Silent"** to enable or disable vibration when the silent mode is enabled.
- Return to the home screen.

How to Set Screen Brightness

- From **Settings,** go to **Display & Brightness.**
- Under the **Brightness** option, click on the indicator and drag either to the left or to the right until you get your desired brightness.
- Click on the >Back sign.
- Click on **General** then **Accessibility.**
- Next, select **"Display Accommodations"**.
- Beside the **Auto-Brightness,** slide the button to the left or right to either switch on or switch off this option.

How to Control Notification Options

- From Settings, go to **Notifications.**
- Click on **Show Preview** and set to **Always** to be able to preview notification on lock screen.
- To set this to only when the device is not locked, click on the option "**When Unlocked**".
- To disable notification preview, select "**Never**".
- Click on the Back arrow at the top left of the screen.

How to Control Notification for Specific Apps

- From the last step above, Click on the specific application.
- On the next screen, beside **Allow Notifications,** move the slide left or right to enable or disable.

How to Control Group Notification
- Scroll down the page and click on **Notification Grouping.**
- Select any of the 3 options as desired.
- Use the Back button to return.

How to Set Do Not Disturb

Your device can be put to silent mode for defined period. Even though your phone is in silent mode, you can set to receive notification from certain callers.

- Under **Settings,** click on **Do Not Disturb.**
- Toggle the switch next to **"Do Not Disturb"** to enable or disable this function.
- Toggle the switch next to **"Scheduled"** then follow instructions on your screen to set the period for the DND.
- Under **Silence** chose **Always** if you want your device to be silent permanently.

- Select **"While iPhone is locked"** if you want to limit this to only when the phone is locked.
- Scroll down and click on **"Allow Calls from"**.
- Chose the best setting that meets your need to set the contacts that can reach you while on DND.
- Click on the back arrow at the top left of the screen.
- Scroll down to **Repeated Calls** and switch the button on or off as needed.
- Click on **"Activate"** under **"Do Not Disturb While Driving"**.
- On the next screen, chose your preferred option.
- Click on the back button to return to the previous screen.
- Scroll down and select **"Auto Reply To"**.

- On the next screen, select the contacts you wish to notify that **Do Not Disturb While Driving** is on.
- Go back to the previous screen.
- Scroll down and select **Auto Reply,** then follow the instructions on the screen to set your auto response message.

How to Turn PIN on or Off
- From **Settings,** click on **Phone.**
- At the bottom of the screen, click on **SIM PIN.**
- Turn the icon beside SIM PIN to the left or right to put off or on.
- Put in your PIN and click on **DONE.** The default PIN for all iPhone XR is 0000.

How to Change Device PIN
- From **Settings,** click on **Phone.**

- At the bottom of the screen, click on **SIM PIN**.
- To change PIN, click on **Change PIN**.
- Type in your current PIN and click DONE.
- On the next screen, type in the new 4-digit PIN and tap DONE.
- The next screen would require you to input the PIN again and click on DONE.

How to Unblock Your PIN

If you enter a wrong PIN 3 consecutive times, it would block the PIN temporarily. Follow the steps to unblock:
- On the home screen, click on **Unlock**.
- Put in the PUK and click on OK.
- Set a new 4-digit PIN and click **OK**.
- Input the PIN again and confirm.

How to Confirm Software Version

- From **Settings,** go to **General** and click on **About.**
- You would see your device version besides **Version** on the next screen.

How to Update Software

- From **Settings,** go to **General** and click on **Software Update.**
- If there is a new update it would show on the next screen.
- Then follow the screen instruction to update the software.

How to Control Flight Mode

- From the top right side of the screen, slide downwards.
- Tap the aero plane sign representing flight mode icon to turn off or on.

How to Choose Night Shift Settings

- From **Settings,** go to **Display & Brightness.**
- Click on **Night Shift.**
- Beside **Scheduled,** click on the indicator and follow the instruction on the screen to select specific period for the Night Shift.

How to Control Automatic Screen Activation

- From **Settings,** go to **Display & Brightness.**
- On the next screen, beside **Raise to Wake,** move the slide left or right to enable or disable.

How to enable Location Services/ GPS on your iPhone XR

- From the Home screen, choose the **Settings** option.

- Scroll towards the bottom of the page and click on **Privacy.**
- Then click on **Location Services.**
- Click on all the apps you would like to have access to your location data.
- Once selected, chose the option **While Using the App.**

How to Turn off location services on iPhone selectively

If there are any apps you would like to block from accessing your location, you can easily turn off location service for such apps by following the steps below.

1. Go to settings on the phone.
2. Move down to the **Privacy** option and then select **Location Services.**
3. You would see all the apps that have access and don't have access to your location. For the apps you wish to access your location information, find such apps,

click on them and select **While Using the App.** For the apps you do not wish to access your location information, find such app, click on it and select **Never.** You can also use these steps for the system services you wish not to grant access to your location information.

How to Turn off location services on iPhone completely

If you do not want any apps or systems on your iPhone to access your location information, follow the steps below to disable it:

- Go to **settings** on the phone.
- Move down to the **Privacy** option and then select **Location Services**
- To turn off the location service, all you need to do is toggle the button and then select **Turn off** to confirm the action. This would prevent all apps and system

services from gaining access to your location data.

How to Use Music Player
- Click on the **Music Player** icon on the home screen.
- Click on **Playlist** then click on **New Playlist.**
- Tap the text box that has **Description,** type in the name for that playlist.
- Click on **Add Music.**
- Go to the category and click on the audio file you want to add.
- Select **Done** at the top of the screen.
- Select **Done** again.
- Go to the playlist and click on the music.
- Use the Volume key to control the volume.
- Click on the song title.
- Tap the right arrow to go to the next music or the left arrow to go to the previous music.
- Gently slide your finger up the screen.

- Click on shuffle to set it on or off.
- Click on Repeat to set it on or off. Here you can select the number of files you want repeated.

How to Navigate from the Notch

Both the sensors and the Face ID cameras are located in the notch found at the top of the device screen. With the notch, you are able to tell the difference between two important gestures which are the notification center and the control center.

1. **Steps to View Notification Center**

You can access the notification center by swiping down from the notch itself or from the left side of the notch.

2. **Steps to View Control Center**

From the right side of the notch, swipe down to view the control center.

Although the notch has occupied most of the space meant for the status bar, however, once you get into the Control center, you would be able to see all the status bar, this includes the percentage of your battery.

How to keep Track of documents

On your iPhone XR, you can access folders and files stored on your iCloud Drive and any other cloud storage services. You can also access and restore folders and files deleted from your device within the last 30 days. There are 3 subsections in the Browse tab and they are:

1. **Locations**: To view files saved in iCloud, simply click on **iCloud Drive**. To view files recently deleted from your device, click on **Recently Deleted**.

   To add an external cloud storage service, you need to first install the app from the App store (Google Drive, Dropbox etc.), then click **Edit** at the right top corner of your device screen to activate it. Once done, click on **Done.**

Other available options for folders and files are:

- To view the content of a folder, click on the folder.
- To Copy, Rename, Duplicate, Delete, Move, Tag, Share or Get info of a folder or file, simply press the folder or file for some seconds.
- To download items with the cloud and arrow icon, tap on them.
- To annotate a file, simply click on the pencil tip icon at the right upper side of the screen. It is important to know that this feature is only available for select image file formats and PDF.

2. **Favorites:** To add folders to the Favorite section, press the folder for some seconds until a menu pop up, then select **Favorite** from the menu. Currently, you can only do

this from the iCloud Device, and only for folders, no single files.

3. **Tags:** when using macros Finder tags, you will see them in the Tags section. Alternatively, press a file for some seconds to tag such file. Then click **"A Tag Here"** to see all the files that have that tag.

How to Move Between Apps

Switching Apps can be tricky without a home button, but the below steps will make it as seamless as possible.

- To open the App Switcher, swipe up from the bottom of the screen and then wait for a second.
- Release your finger once the app thumbnails appear.
- To flip through the open apps, swipe either left or right and select the app you want.

How to Force Close Apps in the iPhone XR

You do this mostly when an app isn't responding.

- Simply swiping up from the bottom of the screen would show the app switcher. This would display all the open apps in card-like views.
- For iOS 12 users, to force close the app, locate the app from the app switcher and swipe up to close the app.
- For users still on iOS 11, press the app you wish to close for some seconds until you see the red button marked with the minus sign at the top of each app card.
- Tap on the minus button for each of the app you wish to close.
- I would advise you upgrade to iOS 12 to enjoy better features on your iPhone XR.
- To go through apps used in the past, swipe horizontally at the bottom of your home screen.

How to Arrange Home Screen Icons

Follow the steps below to arrange the homes screen icon on your iPhone XR.

- Press and hold any icon until all the icons begin to wiggle.
- Drag the icons into your desired position.
- Tap either the **Done** button at the right upper side of the screen or swipe up to exit the wiggle mode.

Complete iPhone XR Reset Guide: How to perform a soft, hard, factory reset or master reset on the iPhone XR

Most minor issues that occur with the iPhone XR can be resolved by restarting the device or doing a soft reset. If the soft reset fails to solve the problem, then you can carry out other resets like the hard reset and master reset. Here, you would learn how to use each of the available reset methods.

How to Restart your/ Soft Reset iPhone

This is by far the commonest solutions to many problems you may encounter on the iPhone XR. It helps to remove minor glitches that affects apps or iOS as well as gives your device a new start. This option doesn't delete any data from your phone so you have your contents intact once the phone comes up. You have two ways to restart your device.

Method 1:

- Hold both side and Volume Down (or Volume Up) at the same time until the slider comes up on the screen.

- Move the slider to the right for the phone to shut down.

- Press the **Side** button until the Apple logo shows on the screen.

- Your iPhone will reboot.

Method 2:

- Go to **Settings** then **General**. Click on **Shut Down.**
- This would automatically shut down the device.
- Wait for some seconds then Hold the **Side** button to start the phone.

How to Hard Reset/ Force Restart an iPhone XR

There are some cases when you would need to force-restart your phone. These are mostly when the screen is frozen and can't be turned off, or the screen is unresponsive. Just like the soft reset, this will not wipe the data on your device. It is important to confirm that the battery isn't the cause of the issue before you begin to fore-restart.

Follow the steps below to force-restart:

- Press the **Volume Up** and quickly release.
- Press the **Volume Down** and quickly release.

- Hold down the Side button until the screen goes blank and then release the button and allow the phone to come on.

How to Factory Reset your iPhone XR (Master Reset)

A factory reset would erase every data stored on your iPhone XR and return the device back to its original form from the stores. Every single data from settings to personal data saved on the phone will be deleted. It is important you create a backup before you go through this process. You can either backup to iCloud or to iTunes. Once you have successfully backed up your data, please follow the steps below to wipe your phone.

- From the **Home** screen, click on **Settings**.
- Click on **General**.
- Select **Reset**.
- Chose the option to **Erase All Content and Settings**.

- When asked, enter your passcode to proceed.
- Click **Erase iPhone** to approve the action.

Depending on the volume of data on your phone, it may take some time for the factory reset to be completed.

Once the reset is done, you may choose to setup with the **iOS Setup Assistant/Wizard** where you can choose to restore data from a previous iOS or proceed to set the device as a fresh one.

How to Use iTunes to Restore the iPhone XR to factory defaults

Another alternative to reset your phone is by using iTunes. To do this, you need a Computer either Mac or Windows that has the most current version of the iOS as well as have installed the iTunes software. Factory reset is advisable as a better solution to major issues that come up from software that wasn't solved by the soft or force restart. Although you would lose data, however, you get more problems fixed including software glitches and bugs.

Follow the guide below once all is set:

- Use the Lightning Cable or USB to connect your device to the computer.
- Open the iTunes app on the computer and allow it to recognize your device.
- Look for and click on your device from the available devices shown in iTunes.
- If needed, chose to back up your phone data to iTunes or iCloud on the computer.
- Once done, tap the **Restore** button to reset your iPhone XR.
- A prompt would pop-up on the screen, click **Restore** to approve your action.
- Allow iTunes to download and install the new software for your device.

How to Choose Network Mode

- From **Settings,** go to **Mobile Data.**
- Select **Mobile Data Options> Enable 4g.**

- To stop using 4g, choose **Off**.
- This option would make your device to automatically switch to either 2g or 3g depending on available coverage.
- Click on **Voice & Data** if you want to use 4g for both mobile data and voice calls.
- Note: To get fast and better connection, use 4g for calls via the mobile network.
- Click on **Data Only** to use for only mobile data.
- Done.

How to set a reminder on iPhone XR

Follow the steps below to create a reminder.

1. Open the **Reminder app** on your iPhone XR device.

2. At the top right corner of the screen, click the plus button to create a new reminder or a list.

3. To create a list, tap **List** and tap **Reminder** to create a new reminder.

4. For reminder, enter the exact reminder content.

5. In the content box space, you have two choices.

   First option: remind me on a day. With this option, please set the Alarm and Repeat options – Every day, Every week, every month, never etc.,

   Second option: Remind me at a location. For this, turn your location on, then set the location you will receive when you arrive or leave.

6. Choose the priority level for the reminder, you can also add notes if needed

7. Chose **Done** to complete the process.

How to set a Recurring Reminder on your iPhone XR

To create a recurring reminder on your device, follow the steps below:

1. Go to the Reminders app on your device.//
2. Type your content on the space for reminder content.
3. Click on the info button beside the new reminder set.
4. Select the option to **"Remind me on a day"**.
5. Set the time you want the reminder.
6. Select **Repeat** and then Custom.
7. Set your frequency to Repeat, Weekly, Daily, Monthly or Yearly.
8. Once done, go to **End Repeat** and select date you want the reminder to stop.

How to get Battery Percentage on iPhone XR

The iPhone XR does not give room to see the battery percentage of your iPhone always, but you can always get a peek to see your battery percentage. To do this, place your finger at the top-right corner of the iPhone XR display and swipe down to be able to access the Control Center. Once the control center is open, you would see the battery percentage at the top right corner of the page.

How to take a Screenshot

Without the home button, taking a snapshot may seem tricky; however, follow these steps to help you take the best shots possible.

- Press both the side and the Volume Up button simultaneously to take a screenshot.
- The photo from the screenshot would be saved automatically in the Photos app,

under the **Screenshots** album. Screenshots help you to note down problems you wish to seek help for later.

- To edit the photo, go to the photo and tap the thumbnail at the left bottom corner of your iPhone.
- To view the screenshots in iOS 11, go to **Photos** click on **Albums** then **Camera Roll/Screenshots**. To do same in iOS 12, go to **Photos,** then **Albums,** go to **Media Types** and select **Screenshots.**

CHAPTER 3: Calls and Contacts

How to Make Calls and Perform Other Features on Your iPhone XR.

This section would give you a detailed guide to contacts and call management in your device.

How to Call a Number

- Tap the Phone icon at the left.
- Click on Keypads to show the keypads.
- Input the number you want to call then press the call icon.
- Tap the end call button at the bottom of the screen once done.

How to Answer Call

- Tap any of the volume keys to silence the call notification when a call comes in.
- If the screen lock is active, slide right to answer the call.
- Click on Accept, if there is no screen lock.

- Tap the end call button at the bottom of the screen once done.

How to Control Call Waiting

- From **Settings,** click on **Phone** then **Call Waiting.**
- Move the icon beside it to the left or right to enable or disable call waiting.

How to Call Voicemail

- Click on the phone icon at the left of the home screen.
- Select **Voicemail** at the bottom right corner of the screen.
- Click on **Call Voicemail** in the middle of the screen and listen for the instructions.
- Tap the end call button at the bottom of the screen once done.

How to Control Call Announcement

Your device can be set to read out the caller's name when there is an incoming call. The contact has to be saved in your address book for this to work.

- From **Settings,** go to **Phone** then **Announce Call.**
- Select **Always** if you want this feature when silent mode is off.
- Choose **Headphones & Car** to activate when your device is connected to a car or a headset.
- The **Headphones Only** option would be for when the device is connected to only headset.
- Select **Never** if you do not wish to turn off this feature.

How to Add, Edit, and Delete Contacts on iPhone XR

Follow the steps below to add, edit and delete contacts on your new device.

How to Add Contacts

- At the **Home** screen, select **Extras**.
- Click on **Contacts**.
- Then select the **Add Contact** icon at the right upper side of your screen.
- Enter the details of your contact including the name, phone number, address, etc.
- Once done inputting the details, tap **Done** and your new contact has been saved.

How to Save Your Voicemail Number

- Once you insert your SIM into your new device, it automatically saves your voicemail number.

How to Merge Similar Contacts
- At the **Home** screen, select **Extras**.
- Click on **Contacts**.
- Click on the contact you want to merge and click on **Edit**.
- At the bottom of the screen, select **Link Contact...**.
- Choose the other contact you want to link.
- Click on **Link** at the top right side of the screen.

How to Copy Contact from Social Media and Email Accounts
- From Settings, go to **Accounts and Password**.
- Click on the account, e.g. Gmail.
- Switch on the option beside **Contacts**.

How to Create New Contacts from Messages On iPhone XR?

- Go to the Messages app.
- Click on the conversation with the sender whose contact you want to add.
- Above the conversation, you would see their phone number.
- Click on the phone number.
- This would show 3 buttons on the screen.
- Click on the **Info** option.
- You would see the number again at the top of the screen, click on it.
- Then click **Create New Contact**.
- Input their name and other details you have on them.
- At the top right hand of the screen, click on **Done**.

How to Add a Caller to your Contact

- On your call log, click on a phone number.

- You would see options to **Message, Call, Create New Contact or Add to Existing Contact.**
- Select **Create New Contact.**
- Enter the caller's name and other information you have.
- At the top right hand of the screen, click on **Done.**

How to Add a contact after dialing the number with the keypad

- Manually type in the numbers on the phone app using the number keys.
- Click on the (+) sign at the left side of the number.
- Click on **Create New Contact.**
- Enter the caller's name and other information you have.
- Or click on **Add to Existing Contact.**

- Find the contact name you want to add the contact to and click on the name.
- At the top right hand of the screen, click on **Done**.

How to Import Contacts

The iPhone XR allows you to import or move your contacts from your phone to the SIM card or SD card for either safekeeping or backup. See the steps below:

- From the Home screen, click on **Settings.**
- Select **Contacts.**
- Chose the option to "**Import SIM contacts**".
- Chose the account you wish to import the contacts into.
- Allow the phone to completely import the contacts to your preferred account or device.

How to Delete contacts

When you remove unwanted contacts from your device, it makes more space available in your internal memory. Follow the steps below.

- From the Home screen, tap on **Phone** to access the phone app.
- Select **Contacts.**
- Click on the contact you want to remove.
- You would see some options, select **Edit.**
- Move down to the bottom of your screen and click on **Delete Contact.**
- You would see a popup next to confirm your action. Click on **Delete Contact** again.
- The deleted contact would disappear from the available Contacts.

How to Manage calls on your iPhone XR

Here, we would talk about how to block calls, set or cancel call forwarding, manage caller ID as well as call logs on your device.

How to Block Calls on the iPhone XR

- Go to **Settings** from the Home screen.
- Click on **Do Not Disturb**. (The DND feature on the iPhone XR allows you determine how you want your device to process incoming calls. The following options are available under DND.

  1. **Do Not Disturb** option – tap on this option to enable or disable the DND feature manually on the device.
  2. Scheduled – To schedule a time for DND to be activate, just tap the tap the time and set the start to end time.
  3. **Allow Calls From** – Use this option only when you want to receive calls from specific people. Select the people and allow calls from them.
  4. **Repeated Calls** – This option allows a call to come through once the call is repeated within 3 minutes of the first call.

How to Block Specific Numbers/Contacts on Your iPhone XR

- Click on the **Phone** icon on the Home Screen.
- Tap **Recent** or **Contacts.**
- Select the specific contact(s) or number(s) you desire to block.
- If accessing through **Recent** option, tap the **(i)** icon next to the number.
- Click on **Block This Caller** at the bottom of the screen.
- Click on the **Block Contact** option to confirm your action.
- Blocked contacts or numbers would be unable to reach you.

How to Unblock Calls or Contacts on your iPhone XR

- Go to **Settings** from Home.

- Select **Phone-> Call Blocking & Identification** then click on **Edit**.
- Click on the **minus (-) sign** next to the contact or number you want to unblock.

How to Use and Manage Call Forwarding on your iPhone XR

With the Call Forwarding unconditional (CFU) feature in the iPhone XR, calls can be forwarded to a separate phone number without the main device ringing. This is most useful when you do not wish to turn off ringer or disregard a call but also do not want to be distracted by such calls. To enable this feature, follow the steps below:

- Go to **Settings** from Home.
- Click on **Phone** then **Call Forwarding**.
- Select the **Forward to** option.
- Input the number you want to forward such calls to.

- You can set the calls to be forwarded to voicemail.

Apart from CFU, Call Forwarding Conditional (CFC) allows you to forward incoming calls to a different number if the call goes unanswered on your number. To enable this feature, you need to have the short codes for call forwarding then set the options to your preference. For data on short codes, reach out to your carrier.

How to Cancel Call Forwarding on your iPhone XR

To cancel,

- Go to **Settings** then **Phone**
- Click on **Call Forwarding.**
- Move the slider to switch off the feature.

How to Manage Caller ID Settings and Call Logs on your iPhone XR

You can decide to hide your caller ID when calling certain numbers. Follow the steps below to activate this.

- Go to **Setting** on the Home screen.
- Click on **Phone** then click on **Show My Caller ID**.
- Click on the switch next to **Show My Caller ID** to either enable or disable the option.

When you disable the feature, the called party will not see your caller ID. This is usually for security or privacy reasons.

How to View and Reset Call Logs on your iPhone XR

For every call you make on your device, there is a log saved on the phone app. To view or manage the call log data, follow the steps below:

- On the Home screen, Click on **Phone** to go to the phone app.
- Click on **Recent** then click on **All.**
- Tap on the call log you wish to extract information from.

How to Reset Call Logs

- Go to **Phones,** then click on **Recent>All>Edit.**
- Click on the **minus (-) sign** to delete calls individually.
- To delete the whole call log once, simply tap **Clear** then chose the **Clear All Recent** option.

CHAPTER 4: Messages and Emails

How to Set up your Device for iMessaging
- From Settings, go to Messages.
- Enable iMessages by moving the slide to the right.

How to Compose and Send iMessage
- From the Message icon, click on the new message option at the top right of the screen.
- Under the "To" field, type in first few letters of the receiver's name.
- Select the receiver from the drop down.
- You would see iMessage in the composition box only if the receiver can receive iMessage.
- Click on the "Text Input Field" and type in your message.
- Click on the send button beside the composed message.

- You would be able to send video clips, pictures, audios and other effects in your iMessage.

How to Set up your Device for SMS

- Your device is automatically set up for SMS once you put in your SIM.

How to Compose and Send SMS

- From the Message icon, click on the new message option at the top right of the screen.
- Under the "To" field, type in first few letters of the receiver's name.
- Select the receiver from the drop down.
- Click on the "Text Input Field" and type in your message.
- Click on the send button beside the composed message.

How to Set up Your Device for MMS

- From **Settings**, go to **Messages**.

- Enable **MMS Messaging** by moving the slide to the right.

How to Compose and Send SMS
- From the Message icon, click on the new message option at the top right of the screen.
- Under the "To" field, type in the first few letters of the receiver's name.
- Select the receiver from the drop down.
- Click on the "Text Input Field" and type in your message.
- Click the Camera icon at the left side of the composed message.
- From Photos, go to the right folder.
- Select the picture you want to send.
- Click Choose and then send.

How to Hide Alerts in Message app on your iPhone XR

- Go to the **Message app** on your iPhone.
- Open the conversation you wish to hide the alert.
- Click on the (i) button at the upper right corner of the page.
- Among the options, one of it is **'Hide alerts'**, move the switch to the right to turn on the option (the switch becomes green).
- Select **'Done'** at the right upper corner of your screen. You are good to go!

How to Set up Your Device for POP3 Email

- From Settings, go to **Accounts and Password**.
- Click on **Add account**.
- Select your service provider from the list or click on others If your service provider is not on the list.
- Select **Add Mail Account**.

- Input your details, name, email address and password.
- Under Description, put in your desired name.
- Click Next at the top right corner of the page.
- The next screen is a confirmation that your email has been set up.
- Follow the on-screen instructions to enter in any extra information.

How to Set up Your Device for IMAP Email

- From Settings, go to **Accounts and Password**.
- Click on **Add account**.
- Select your service provider from the list or click on others If your service provider is not on the list.
- Select **Add Mail Account.**
- Input your details, name, email address and password.

- Under Description, put in your desired name.
- Click Next at the top right corner of the page.
- The next screen is a confirmation that your email has been set up.
- Follow the on-screen instructions to enter in any extra information.
- After this, select **IMAP,**
- Under host name, type in the name of your email provider's incoming server.
- Fill in the username and password for your account.
- Under outgoing host server, type in the name of your email provider's outgoing server.
- Click **Next.**
- Select **Save** at the top right of the screen to save your email address.

How to Set up Your Device for Exchange Email

- From Settings, go to **Accounts and Password.**
- Click on **Add account.**
- Select **Exchange** as your email service provider.
- Input your email address.
- Under Description, put in your desired name.
- Click on **Sign In.**
- Input your email password on the next screen.
- Click on **Sign In.**
- Move the indicator next to the needed data type to enable or disable data synchronization.
- Select **Save** at the top right of the screen to save your email address.

How to Create Default Email Account
- From Settings, go to **Mail** at the bottom of the page.
- Click on **Default Account.**
- On the next screen, click on the email address you wish to set as default.

How to Delete Email Account
- From Settings, go to **Accounts and Password.**
- Click on the email address you want to delete.
- Select **Delete Account** at the bottom of the page.
- On the next screen, click on **Delete from my iPhone.**

How to Compose and Send Email
- From the Home screen, select the Mail icon.
- Click on the back arrow at the top left of the screen.

- Select the email address you want to send the email from.
- Click the new email icon at the bottom right side of the screen.
- On the To field, input the receiver email address and the subject of the email.
- Write your email content in the body of the email.
- To insert a video or picture, press and hold the text input field until a pop-up menu comes up on the screen.
- Click **Insert Pictures or Videos** from the pop-up and then follow the instructions you see on the screen to attach the media.
- To attach a document, select **"Add Attachment"** and follow the instructions you see on the screen.
- Click on **Send** at the right top of the screen.

CHAPTER 5: Manage Applications and Data

How to Install Apps from App Store

- Open the app store and click on search.
- Type in the name of the app in the search field.
- Click on Search.
- Select the desired app.
- Click on **GET** beside the app and follow the steps on the screen to install the app. For paid apps, click on the price to install.

How to Uninstall an App

To uninstall an app,

- click and hold the app until it begins to shake.
- Click on the **Delete** option, then select **Delete**.

With this method, every settings and data about the app would be deleted from your phone.

How to Delete Apps Without Losing the App Data

- From the **Settings,** go to **General.**
- Click **iPhone Storage.**
- Click on the app you wish to uninstall and click on **Offload App.**
- Select **Offload App** again to complete.

How to Control Offload Unused Apps

You can set your device to uninstall apps that are not used in a long time. The app would be uninstalled without deleting the data from the phone. Follow the steps below:

- From the **Settings,** go to **iTunes and App Store.**
- At the bottom of the screen, beside **"Offload Unused Apps",** move the switch left or right to control it.

How to Control Bluetooth

- From **Settings,** go to **Bluetooth**
- Move the switch beside **Bluetooth** to switch on or switch off Bluetooth.
- To pair with a mobile device, put on the Bluetooth then click on the device you want to pair and follow the steps on the screen to link.

How to Control Automatic App Update

- From the **Settings,** go to **iTunes and App Store.**
- Beside **"Update" option,** move the switch left or right to control it.
- Move to **"Use Mobile Data",** move the switch left or right to enable or disable.

How to Choose Settings for Background Refresh of Apps

- From the **Settings,** go to **General.**
- Click on **Background App Refresh.**
- Then click on **Background App Refresh** again.
- Select **OFF** to disable.
- To refresh the apps using Wi-fi, select **Wi-fi.**
- Select **Wi-fi and Mobile Data** if you want to be able to refresh using mobile data.
- Use the back button to return to the previous screen.
- For each of the apps listed, move the slide either left or right to enable or disable.

How to configure your iPhone XR for manual syncing

- Using either Wi-fi or USB, let your device be connected to a computer.
- Manually open the iTunes app if it doesn't come up automatically.
- Tap on the iPhone icon on the top- left of the iTunes screen. If you have multiple iDevice, rather than seeing the iPhone icon, you would see menu showing all the connected iDevices. Once the devices are displayed, select your current device.
- Tap on the **Apply** button at the bottom right corner of your screen.
- Tap on the Sync button if it doesn't start syncing automatically.

How to Synchronize using iCloud
- Click on your Apple ID under Settings.
- Click on iCloud.
- Scroll down to **iCloud Drive** and move the switch left or right to enable or disable.
- Under iCloud, click on **Photos.**
- Scroll down to **Upload to My Photo Stream** and slide left to right to activate or disable.

How to manually add or remove music and videos to your iPhone XR

To manually manage your music and videos, you would have to copy the video files and music tracks to the iPhone from the iTunes Library. Follow the steps below to do this:

- Connect your device to your computer or Mac.
- Launch the iTunes app.
- Manually move the media to the left side of the window.

- Release the media on top of the iPhone (Under Devices).
- Now you can drag any of the items from the main window to the sidebar to add to your iPhone from iTunes.

How to Choose Settings for Find my iPhone
- Click on your Apple ID under Settings.
- Click on iCloud.
- Scroll down and click on **Find My iPhone.**
- Slide left to right to activate or disable.
- Scroll down to **"Send Last Location"** and Slide left to right to activate or disable.

How to Use Find My iPhone

This option helps to recover your iPhone when lost. To activate:

- Open a browser on your computer and go to www.icloud.com
- Select **Find iPhone.**

- At the top middle of the screen, select **All Devices.**
- Select the name of your mobile device from the drop down.
- The next screen would show you your device latest position on the map. (Ensure that you have activated sending of your mobile phone's latest position.)
- Tap **Play Sound.** This would send a signal to your lost device that would play back for 2 minutes. For the play back to happen, your device has to be connected to a strong connection.
- To lock your device, click on **Lost Mode** on the screen and follow the steps to lock the device.
- Not only can you lock your iPhone with a code, you can also set up a lost message to show on the screen of the device.
- To delete all the phone content, select **Erase iPhone.** Once this is done, you

would be unable to use Find my iPhone for that device.

How to Downgrade iOS System on Your iPhone

Did you just recently upgrade your iOS but want to go back to the previous iOS you are familiar with? Here, you will learn tips on how to downgrade without loss of data.

1. First, it is important you back up your device data. I would advise you backup using iCloud as backing up with iTunes can affect your device system when you restore. This would make it easy to restore from iCloud once you have downgraded the iPhone.

To backup, follow the steps below:
- Go to **Settings>iCloud.**
- Look out for the button titled **'Backup" or "iCloud Backup",** switch it on.
- Ensure your device is connected to Wi-fi and device must be charging during the process.

- Once the backup is done, visit **Settings>Name>iCloud>iCloud Storage>Manage Storage** to confirm the phone backup.
2. Now you are ready to downgrade the system. To downgrade, you need to have a backup file from the iOS you want to switch to. If you don't have any, you can get a standard file downloaded from **Apple Support** to the iOS system. To do this,
- Visit Apple Support, navigate to the Download page
- In the product list, find and select iPhone, then select your desired iOS system.
- Select the option to download to your PC.
- While the file is downloading, update the iPhone on your computer to the most current version.
- Use the lighting cable to connect your phone to your PC.

- On the top left, click the iPhone icon and switch mode to **iPhone Device Panel.**
- Look for the "Restore Backup" button and select it.
- Select the file gotten from the Apple support to downgrade your iPhone to your choice system.
- Once the downgrade is done, you can visit **Setting>General>Software Update** to confirm that the downgrade was done successfully.

3. Once you are done with the downgrade, next is to restore the data saved. Do not worry if you cannot find the data on your iPhone. Just go to the iCloud Backup to restore them. To do this, follow the steps below

- Go to **Settings>General>Reset.**
- Here, you are able to reset the needed data and also "**Erase All Content and Settings**" directly.

- On the App and Data screen, click on "Restore from iCloud"
- Next, input your login details into the iCloud account to choose the backup file you wish to restore.
- Allow it to restore without interruption so that you can get all the files you lost after downgrading.
- That is all there is to downgrading.

CHAPTER 6: Internet and Data

How to Set up your Device for Internet
- Your iPhone is automatically set for internet once the SIM Card is inserted.

How to Use Internet Browser
- Click on the internet browser icon.
- Go to the address bar at the top of the page and input the web address. Then tap Go.
- Click on the menu icon at the bottom of the screen.
- Click on **Add Bookmark.**
- Under Location, click on Favorites and click on Bookmarks.
- Type in the name for the page you want to save and click on save.
- Tap the bookmark icon next to the menu icon.
- Click on the website under bookmark you want to visit.

How to Clear Browser Data

- Go to Settings, then click on **Safari**.
- On the next screen, click on **Clear History and Website Data**.
- From the pop-up, click on **Clear History and Data**.

How to Check Data Usage

- Go to **Mobile Data** under **Settings**.
- Beside **Current Period**, you would see your data usage on the device.
- Under each app, you would see the data usage for those apps.

How to Control Mobile Data

- Go to **Mobile Data** under **Settings**.
- Move the switch beside **Mobile Data** to the right or left to put off or on.
- Scroll to where you have the applications and move the switch beside each app to the right or left to put off or on.

How to Control Data Roaming

- Go to **Mobile Data** under **Settings**.
- Click on **Mobile Data Options**.
- Move the switch beside **Data Roaming** to the right or left to put off or on.

How to Control Wi-fi Setup

- From the top right side of the screen, draw down the screen.
- Click on the Wi-fi icon to enable or disable.
- Move the switch beside **Wi-fi** to the right or left to put off or on.

How to Join a Wi-fi Network

- Go to Settings, then click on Wi-fi.
- Move the switch beside **Wi-fi** to the right to put on the Wi-fi.
- Select your Wi-fi network from the drop down.
- Type in the password and click on **Join**.

How to use your iPhone as a Hotspot

- Go to **Personal Hotspot** under **Settings**.
- Move the switch beside **Personal Hotspot** to the right or left to put off or on.
- IF wi-fi is disabled, click **Turn on Wi-fi and Bluetooth**.
- Select **Wi-fi and USB only** if wi-fi is enabled already.
- Input the wi-fi password beside the field for wi-fi password.
- Select **Done** at the top of the screen.

How to Control Automatic Use of Mobile Data

- Go to **Mobile Data** under **Settings**.
- Move the switch beside **Wi-fi Assist** to the right or left to put off or on.

CHAPTER 7: What is iCloud Backup and How to Use it

The word **"iCloud backup"** is used frequently but not everyone know what it is and the correct way to use it. Here we would talk on iCloud backup and its use.

What is iCloud Backup?

iCloud is a limited online space that Apple offers to all its users. This means that for every one that have an Apple account, they can enjoy the benefits that comes with using iCloud. iCloud would help you as an Apple user to sync all your Apple devices. Then, what is iCloud backup. iCloud backup allows you backup all your apple devices as well as computers on the iCloud. And then you can move data backed on the iCloud to any of your iPhones whenever you need it.

What files can you back up on iCloud

- Messages (plus SMS, iMessage, MMS)

- App data
- Visual voicemail
- Call history
- App layout and home Screen
- Ringtones
- HomeKit configuration
- Settings
- Videos and photos
- History of all purchases done on Apple.
- Apple Watch backup

Data like bookmarks, contacts, notes, mails, iCloud photo library, calendar, shared phots, My Photo Stream and other data stored on the iCloud cannot be backed up because they are already stored in the iCloud.

How to sign into iCloud on your iPhone XR.
- Go to the **Settings app.**
- At the top of your screen, click on **Sign in to your iPhone.**

- Enter your Apple ID email address and password.
- Then click on **Sign In.**
- Next screen would ask for your device passcode if you set up one.
- Set the iCloud Photos the way up like them.
- Switch **Apps using iCloud** on or off, however you want it.

How to Sign Out of iCloud on Your iPhone XR
- From the **Settings app,** click on **Apple ID.**
- Click on **Sign Out** at the bottom of the screen.
- Input your Apple ID password then select **Turn Off.**
- Chose the data you would like to keep a copy of on your iPhone and move the switch on.
- At the top right corner of your screen, click on **Sign Out.**

- Click on **Sign Out** to confirm your decision.

How to Use iCloud Backup

To use the iCloud backup, ensure that your device is connected to Wi-fi before you proceed. To back up, follow the steps below

- Navigate to **Settings>Name>iCloud**. If you are using other iOS, go to **Settings>iCloud**.
- Look out for the button titled **'Backup" or "iCloud Backup"**, switch it on.
- Ensure your device is connected to Wi-fi and device must be charging during the process.
- Once the backup is done, visit **Settings>Name>iCloud>iCloud Storage>Manage Storage** to confirm the phone backup.

How to Troubleshoot if iCloud isn't Working

If your iCloud isn't working, follow the steps below:

- Ensure the Wi-fi is connected and strong as this is usually the main reason for iCloud backup to not respond.

- Once done, confirm that you have enough space in the cloud. Apple provides only 5G free. If you have used up the spaces, clear the files you don't need or rather back them up with iTunes then remove them from the iCloud backup.

- If you do not wish to delete any information, next step would be to purchase additional room in the iCloud. Please see the pricing below.

  50 GB per month: 0.99 USD

  200GB per month: 2.99 USD

  2 TB per month: 9.99 USD

- Lastly, remove any unwanted data from the iPhone or computer before you perform the iCloud backup.

How to share a calendar on iPhone XR via iCloud

To share your calendar on your iPhone, it is important to first of all turn on the iCloud for calendar option. Kindly follow the steps below:

- On your iPhone, go to **'settings'**
- Click on your device name and select **"iCloud"**
- Then turn on **"Calendars"**

After this is done, you can now share your calendar by following these steps

- Open the **"Calendar"** app on your device.
- At the bottom of your screen, select **"calendars"**.
- You would see an **"info"** icon next to the calendar you want to share, click on the icon.

1. Select the **'add person"** option on the screen then pick the people you wish to share the calendar with.
2. Tap "add" followed by "Done" at the top of your screen.

CHAPTER 8: Troubleshooting the iPhone XR Device

Most challenges encountered with the iPhone XR can easily be resolved by restarting your device. However, in this section, we would look at every possible challenge you may have with the iPhone XR and the solutions.

Troubleshooting Basic Functions

How to Fix "My iPhone is not coming on"

There are 3 possible reasons for this and their solutions:

- **Battery is damaged:** Get a new battery.

- **Battery is drained:** Charge the battery.

- Your mobile phone didn't start up correctly: Press and hold the Side button until the phone comes up. Swipe up the screen from the bottom. Input your Sim PIN if it is locked. The default password is 0000. If you put in an incorrect PIN thrice, your SIM gets locked and you would need your PUK key to unlock. You can get this from your carrier's customer care. It is important to know that if you put in the wrong PUK number 10 times, it would automatically block the SIM.

I Can't Activate my Mobile Phone

There are four possible reasons for this. Reasons and solutions are listed below:

- **Activation is not properly done:** follow the activation steps listed in this book.

- **There is a temporary problem with the network:** you can attempt activating at a later time or you change the area where you at.

- **The problem is with the SIM:** Contact your carrier's customer service for this.

- **There is no adequate network coverage:** you can attempt activating at a later time or you change the area where you at.

My Mobile Phone Doesn't Respond

1. **There may be an issue with the iOS:** restart the device.

Screen lock is on: disable screen lock. Follow the steps below to disable screen lock.

- Press the **Side button.**
- Slide your fingers up the screen.

To set automatic screen lock:

- From **Settings,** go to **Display & Brightness.**
- Click on **Auto-Lock** and chose your preferred settings.

My Device Memory is Full

This is probably because your applications are taking too much space. To resolve it, simply uninstall apps you do not need. To uninstall an app, click and hold the app until it begins to shake. Click on the **Delete** option, then select **Delete.**

How to Delete Apps Without Losing the App Data

- From the **Settings,** go to **General.**
- Click **iPhone Storage.**
- Click on the app you wish to uninstall and click on **Offload App.**
- Select **Offload App** again to complete.

My Mobile Phone is Slow

Possible reasons and their solutions are:

- **You have too many applications running at same time**: close some of the apps that are running (follow the steps listed in the book).

- **Your device is overloaded:** restart your device.

My Device's Battery Life is Short

There are 14 possible reasons for this as listed below:

- **Live wallpapers are enabled:** turn it off

- **Auto screen lock is disabled:** Enable automatic screen lock. Follow the steps already discussed in this book to do this.

- **Auto app update is on:** turn off this feature. Go to **Settings>iTunes & App Store.** Move the slider beside the **"Updates"** option to the left or right to switch on or off. Move the slider beside the **"Use Mobile Data"** option to the left or right to switch on or off. Return to home screen.

- **Auto content sync is enabled:** disable this option.

- **Battery is damaged:** get a new battery.

- **Bluetooth is active:** switch off the Bluetooth using the steps already discussed.

- **Mobile data is on:** turn off mobile data.

- **Notifications are enabled**: turn it off

- **Screen brightness is too high:** Reduce the screen brightness

- **Vibration is on:** Disable vibration

- **Wi-Fi is on:** Turn off Wi-fi

- **Several applications are running:** Close the applications you are not using.

- **GPS is enabled:** Disable GPS.

No Ringtone is Heard on Incoming Calls

Ensure that you can make call first, and also confirm that the Ring volume is on and not muted. If this is confirmed, you should be fine.

What to do When You can't make Voice Call

The possible reasons and solutions to this are listed below:

- **There may be a problem with the receiver:** Try calling another number to see if it will connect.

- **This could also be due to poor network:** You can postpone the call till when there is better network.

- **Absence of network coverage:** You can postpone the call till when there is better network.

- **You do not have sufficient airtime on your prepaid line:** Top up your credit.

- **Your device is overloaded:** Restart your device.

- **Flight mode is activated:** Deactivate the flight mode.

- **Your number has been suspended:** Contact customer service to activate reactivate your line.

- **Your SIM is bad:** Contact customer service for replacement.

- **The selected network is not in range:** Activate the automatic network selection option.

Calls and Voicemail Troubleshooting

How to Fix "I am not Receiving Messages on my Voicemail"

Reasons and possible solutions are below:

- **Your voicemail is not active.** Call your carrier on how to activate the voicemail.

- **You haven't set up calls to the voicemail:** Set up calls to be diverted to voicemail.

How to Fix "I Can't Listen to my Voicemail"

- **Your Voicemail may not have been set up:** Set up your voice mail.

- **You are attempting to listen to the voicemail from another mobile:** Contact your carrier to know how to listen to voicemail from another device.

How to Fix "I Can't Receive Voice Calls"

- **This may be a connection problem:** Simply restart the device.

- **You may have activated Divert All Calls option:** Just disable the option.

- **It could be that the device diverts missed calls too quickly:** Simply set up **Divert Delay.**

Messages and Email Troubleshooting

I am Unable to Send or Receive Messages

Before any other solution, first confirm that you can make calls. If yes, then the possible problems and solutions are below.

- **Sending of iMessage is Enabled:** Simple turn it off.
- **The error could be from the receiving end:** try sending to a different number.

I am Unable to Send or Receive MMS

- **Data Roaming is disabled:** Turn it on.
- **Problem is from the receiver:** Try sending to a different number.

- **Your mobile phone is not rightly set up for MMS:** follow the right steps stated in this book.

- **Mobile data is turned off:** turn it on.

I am Unable to Send or Receive iMessages

- **Problem is from the receiver:** Try sending to a different number.

- **There may be a problem with activating iMessage:** Turn off iMessage and activate afresh.

- **Your phone is not properly set up for iMessages:** Follow the steps in this book to set up iMessage.

- **Time zone, Date and Time are not Correct:** follow the steps given to set up date and time.

Entertainment and Multimedia Troubleshooting
I Can't Install Apps

Before any other step, first confirm that the phone is connected to a working internet network. If network is good, then below are the possible problems and solutions.

- **Wrong Apple ID payment information inputted:** Log into the app store from your computer and input an updated payment information.

- **The app you desire is not available in your country:** wait till its available or search for similar apps.

- **This could be due to insufficient space:** delete aps that you are not using.

- **Your Apple ID is not active on your device:** Activate the Apple ID on your iPhone XR.

How to fix "iPhone won't download and update App" with 5 tips

Confirm the network connection on your iPhone:

If your network connection is unstable or bad, its very likely that the download of apps would pause due to the bad connection. If this is the case, all you need do is change to a faster network or leave the app to download at a better time.

Free up space on your iPhone

If your network is good, then you need to check that you have available space to contain the downloads and updates. If there are no space, the app would be unable to download or install successfully. To get your device storage, do the following

- Under the "settings" app, select "General"
- Then select "Storage and iCloud Usage".
- Here you can see space used. If all the space has been used up, you can make some space by deleting useless and unwanted files. Once this is done, attempt to download or install afresh.

RE-install app

Another step you can take to fix the download or update issue is to delete the app and download the latest version afresh. Then install it. Before you delete, ensure you back up any needed data.

Check your Apple ID

Confirm that you are successfully signed into the Apple store with the correct details that you use for purchase, for users that have multiple Apple ID. Alternatively, logout and login afresh to the store.

Try to update the Apps with iTunes

If you have tried to update on iPhone without success, maybe you should try updating via iTunes instead. Follow the steps below:

- Connect your device to your computer and start the iTunes
- At the left-hand side of the iTunes page, you have a drop-down button, select the "Apps" option. Then select Updates at the middle top of the page.

- You have two options to update the apps. Either you select "Update all apps" at the bottom right of the page to update all the apps, or you right click on individual apps and select "update App" to update each app.

Note that if the error is caused by Apple store server, the only thing you can do is to wait for Apple to fix their issues.

I Can't Play Music on my Device

Possible reasons and solutions are below:

The music player does not support the audio file: Play an audio file supported by the device.

There is no audio file on your iPhone: move audio from your computer to the phone.

- **The headset is damaged:** get a new headset.
- **The audio file is bad:** delete the file and send it again from your computer.

I Can't Use GPS Navigation

First confirm you have an active working internet connection.

- **Poor GPS Signal:** go somewhere with a good view of the sky and attempt to search for signal.
- **GPS is not enabled:** Put on your GPS.

I Can't Take Pictures with the Device Camera

You do not have enough space: delete some apps and delete unwanted items from your iPhone.

Fixes to Photos Not Showing Up on iPhone

If you are unable to see your new photos on the iPhone Camera roll, follow the two tips below for help.

Tip 1: Restart iPhone

1. Restart your iPhone XR.

2. Once the phone is up, take a photo with your phone.

3. In the camera app, click the thumbnail of the photo you took to view the picture.

4. To send this photo via iMessage, tap the **Share** button.

5. Restart the iPhone and then go to the **Photos** app on your iPhone to check for the picture.

Tip 2: Update to the latest version of iOS

It is possible that you are not seeing the photos in the Camera Roll due to a bug in the iOS of your device. To solve this, upgrade to the latest version if you are not already using the most recent version. To do this, visit **Settings > General > Software Update**. You can also explore this tip if your FaceTime live photos are not saving.

Part 2: Solutions to Photos Not Showing Up on iPhone after restore

Tip 1: Perform a restart on iPhone

This is by far the commonest solutions to many problems you may encounter on the iPhone XR. You have two ways to restart your device.

Method 1: Press and hold both the **Side** button and **Volume Up** (or **Volume Down**) at same time until you see a slider on the screen, move the slider for the phone to go off completely and then turn on your device by pressing same buttons.

Method 2: Go to Settings > General > Shut Down, then you see a slider on your screen, move the slider for the phone to go off completely. Hold the **Side** button to power on your iPhone.

Tip 2: Turn on iCloud Photo Library

Did you just restore from iCloud backup and can't find your photos? Its important you know that turning on the iCloud photo library when backing up your iPhone to iCloud means the photos won't be among the items in the iCloud backup. To keep the photos on your restored iPhone, go to **Settings > [name on device] > iCloud > Photos,** then switch on **iCloud Photo Library** before you connect your device to a good Wi-fi connection and allow the iCloud photos to download into your iPhone.

If you did not turn on iCloud Photo Library, all your photos should be included in the backup. To restore back, ensure you have a good Wi-fi connection, then wait patiently for the photos to download into your iPhone.

Part 3: Fixes to iCloud Photos Not Showing Up on New iPhone

Did you move the photos on your old iPhone to iCloud but not able to move the photos to your new device? If you are unable to see the iCloud photos on your new device, simply follow the steps below:

- Check iCloud settings

Turn on the iCloud Photo Library before you begin to download the photos to your iPhone. To check the settings, follow the steps below:

1. Open the **Settings** app on your iPhone.

2. Tap [name on the device] > iCloud > Photos.

3. Turn on **iCloud Photo Library**.

4. Ensure your iPhone is connected to a good and stable Wi-Fi connection before you begin to download the photos.

- **Check the Apple ID on iPhone**

You have to use the Apple ID linked with the old device to sign into the new iPhone. If using the correct one, just sign out of iCloud and sign in again.

- **Check your network connection**

You need a stable Wi-fi connection to be able to sync photos from iCloud to your iPhone. You can change the network connection if your iPhone is connected to a weak connection.

Connectivity Troubleshooting

I Can't Use my Device Internet Connection

- **Data Roaming is disable:** Enable it.
- **The error could be from your device iOS:** restart your phone.
- **Mobile data is off:** enable mobile data.

I Can't Use Wi-Fi

- **There is no established Wi-fi connection:** connect to a working wi-fi network.
- **There are no Wi-fi networks available:** connect to a working wi-fi network.
- **The connection is rejected by the Wi-fi you choose:** Reach out to the wi-fi admin for the correct connection settings.

- **Although your device is connected to a wi-fi, yet you are not logged on:** for some connections, you would need to log on first before use. Simply go to the browser and attempt to open any webpage. The page would direct you to start page of the wi-fi network. Login using the on-screen instructions.
- **Wi-fi is turned off:** Put on Wi-fi.

How to fix Bluetooth Not Working on iPhone XR

- Confirm that the Bluetooth accessory is among the list of supported devices on your iPhone. If the Bluetooth device isn't supported, you would be unable to connect it to your iPhone.
- Ensure that both your iPhone and the Bluetooth device have sufficient power supply. If either of the two devices have low battery, you would be unable to connect the iPhone to the Bluetooth.

- Check to confirm that the Bluetooth on your iPhone and the Bluetooth accessory is on. To check that this on your iPhone, go to **Settings** then select **Bluetooth.** If the Bluetooth is not coming on, simply restart your iPhone.
- Place both the iPhone and the Bluetooth accessory beside each other. The connection would likely not succeed if the distance between the two devices goes beyond the supported distance.
- Put on both the Bluetooth accessory and the iPhone Bluetooth.
- If the connection is still unstable, just disconnect the pairing, that is, unpair both devices and re-pair them again.
- Restart both the Bluetooth accessory and your iPhone.
- If all the steps above fail, you would need to reach out to Apple support.

My Phone Uses a lot of Mobile Data

Possible problems and solutions are:

- **GPS position usage for apps is on:** Disable "GPS Position" for specific applications.

- **Mobile data usage for app is enabled:** disable this feature.

- **Auto mobile data usage is enabled:** disable this option.

- **Auto update of apps using mobile data is enabled:** disable this option

- **Notifications are enabled:** turn it off or disable it for select apps.

- **Background refresh of app is enabled:** disable this option.

How to Fix iPhone Red Screen

Some iPhone XR users have complained of getting a red screen when starting their device and then the phone just keeps restarting. I know that this can be very frustrating but do not worry as I will show you some solutions to this.

1. Fix the Red Screen by Restarting your iPhone

The simplest solution to this problem is to restart the iPhone with the steps below:
- Hold the Power button until you see "Slide to power off" on your screen.
- Swipe left to right to turn off your iPhone.
- Press and hold the Power button again to turn on iPhone
- After restarting, you can check if the screen will turn to red again.

2. Fix iPhone Red Screen by Resetting the iPhone

We have talked about the steps to hard reset or force restart your iPhone. Simply follow that steps here.

- **Fix iPhone Red Screen by Putting the iPhone XR in DFU Mode**

If all the above methods didn't help resolve the red screen issue, then try putting your device into the DFU mode using iTunes. Before you do this, ensure to back up your data as this process would wipe out data from the iPhone XR.

- On your computer, run the iTunes App.

- Connect your device to the computer.

- Put off the iPhone XR.

- In quick intercessions, press the **Side** button for about 3 seconds and then hold down the Volume Down button for 10 seconds, after which you release the Side button while still holding down the Volume Down button until the screen goes completely black.

- Once the screen is black means the device has gone into DFU mode.

CHAPTER 9: Conclusion

Now that you have known all there is to know in the iPhone XR, I am confident that you would enjoy operating your device.

The iPhone XR has helped to make things easy and reduce stress only if you have the right knowledge and know how to apply it which I have outlined in this book.

All relevant areas concerning the usage of the iPhone XR from taking out of the box to setup and operations has been carefully outlined and discussed in details to make users more familiar with its operations as well as other information not contained elsewhere.

If you are pleased with the content of this book, don't forget to recommend this book to a friend.

Thank you.